PENGUIN HANDBOOKS

THE AVOCADO PIP GROWER'S HANDBOOK

Hazel Perper lived in New York City, where she was known far and wide as the 'avocado lady'. The widow of artist Mark Perper, she was at one time deeply involved in the art world, and traced her interest in raising plants from pips, seeds and cuttings back to the time of the Depression, when she first began 'making do'.

In addition to *The Avocado Pip Grower's Handbook*, she also wrote *The Citrus Seed Grower's Indoor How-To Book*, and *Indoor How-To Book of Oats, Peas, Beans and Other Pretty Plants*.

Hazel Perper died in 1977.

Hazel Perper

The Avocado Pip Grower's Handbook

Illustrated by Timothy Perper

PENGUIN BOOKS

Penguin Books Ltd, Harmondsworth,
Middlesex, England
Penguin Books, 625 Madison Avenue,
New York, New York 10022, U.S.A.
Penguin Books Australia Ltd, Ringwood,
Victoria, Australia
Penguin Books Canada Ltd, 2801 John Street,
Markham, Ontario, Canada L3R 1B4
Penguin Books (N.Z.) Ltd, 182–190 Wairau Road,
Auckland 10, New Zealand

First published in the U.S.A. by Walker
and Company 1965
Published simultaneously in Canada by
George J. McLeod Ltd, Toronto, 1965

This revised edition first published in
Great Britain by Penguin Books 1980

Made and printed in Great Britain by Butler and Tanner Ltd,
Frome, Somerset
Set in Photina by Filmtype Services Limited, Scarborough

Contents

Preface

Anyone who enters my living-room for the first time inevitably says, 'It's beautiful! What is it?' It's an indoor avocado plant. It is ceiling high, with a spread of six feet. Some of its glossy, dark green leaves are fifteen inches long. Two and a half years ago it was the pip left over from a delicious lunch.

Writing a handbook on indoor avocado growing is a self-defensive action intended to liberate me from a number of small harassments – from the midnight telephone call ('We've just done the dishes and we're ready to toothpick the avocado, but I forgot whether you said to *wash* it or not') to the unexpected coffee break with an upstairs tenant ('You don't know me, but I'm from the flat above and a neighbour told me about you: so *you're* the avocado woman').

Bad enough to be a sort of botanical midwife, called out of the night at odd hours on pointers on toothpicking a pip. Worse still to hear oneself referred to as an Avocado Woman. But this situation becomes nearly intolerable when everybody who comes to our home talks about nothing but avocados. The avocado is instantly the first and exclusive subject of conversation whenever a group gathers in my living-room. If there is a newcomer present, I know that I will have to spend a good part of the evening talking about the tree and its care and growth. I can predict to a nicety the way the conversation will go, rating the probable incidence of comments, questions and so forth about as follows:

'That's an *avocado*?!' A certainty.

'How did you do it?' Excellent chance.

'How long did it take?' Very good chance.

'I once tried one but it grew straight up and didn't have many leaves. What did I do wrong?' Fair chance.

'What kind of soil?' Only a possibility.

'Will it grow fruit?' Absolute certainty.

And when answers are offered, or when I try to outline an overall theory:

'You have to cut back the first stem.'

'*What?* The first *stem*, you mean cut it off?'

'And keep cutting back.'

'Really?' An expression of surprise and a dubious glance, followed by a walk to the tree and a close look into the foliage. 'Hmmmm.'

'It's a tropical tree and has to be kept warm.'

'You don't say.'

'Never use cold water.'

'Is *that* it? I never knew that.' And, at times, an envious note is sounded. 'You must have a green thumb.'

A vehement denial on my part, but a chance, at last, to change the subject.

After months of being tyrannized by my conversation piece, I decided life was far too full of a number of other things to give so much time to the repetition of the same formula. *Voilà: 'The Avocado Pip Grower's Handbook.'*

Though I hope the emergency phone calls are a thing of the past, I plan to present copies of this book to friends who might otherwise feel abandoned in mid-plant. For, all rebelliousness aside, I do not forget how attached one can become to something green that one has grown, by oneself, indoors, and all the way up from a mere seed. It's a peculiarly satisfying experience.

1

The Avocados Available, and Their Sources

Many people have tried, at one time or another, to grow an avocado tree indoors. The instructions they follow are usually word-of-mouth, and often something important is left out. They start the pip, or 'pit', to name it correctly, upside-down in a glass of water; they get discouraged when nothing happens *immediately*; they don't know how to plant the pip; they don't know how, when or why to cut back the stem.

Avocado-growing is really very simple. It takes little of your time. And you grow a stunning plant – an indoor tree for which florists could charge an exorbitant price.

Every stage of the growth is fascinating to watch: from first roots and first shoots to the first tiny leaves, held tight and closed at the tip of the stalk, to the gradual production of leaves by the fully grown plant. The avocado is hardy and cooperative. You can control its size – if you don't want a tree, you can

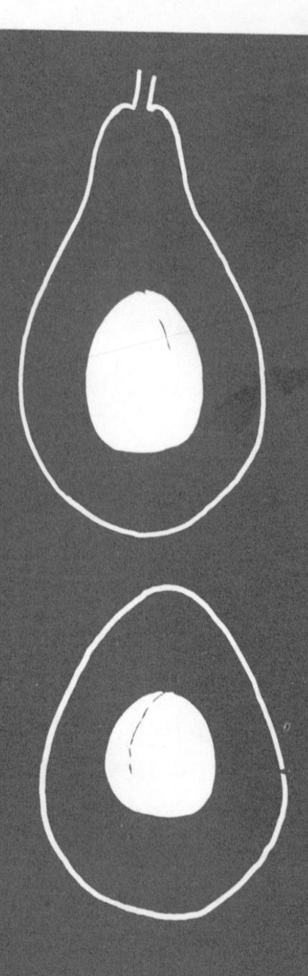

TWO VARIETIES OF AVOCADO:

THE WEST INDIAN (TOP)

AND THE GUATEMALAN

keep it plant-size. It will last indefinitely. I've had many of them for years and have only lost them by giving them away.

Many different kinds of avocados are sold in British markets. As a result of extensive hybridization they vary widely in shape, weight, colour and flavour. Although avocados are grown commercially in California, Florida, South Africa, Australia, Brazil, Cuba, the Pacific Islands and Israel, most of the fruit sold in Britain comes from Israel.

There are two main varieties of avocados: West Indian and Guatemalan. The West Indian, which is less hardy (more susceptible to frost) grows in Florida, while the Guatemalan type is grown widely in California, so sometimes these varieties are called 'Florida' or 'California' avocados.

West Indian

This is the larger fruit and, in general, the best and easiest to work with. It may vary in weight from ten ounces to a jumbo forty ounces. The colour is usually fairly dark green with a brownish tint and an almost-smooth texture. Or some – generally the larger ones – may be a dusky purple, with a pebbled and roughly mottled surface. The flesh is smooth and soft, high in oil content. The pip is not as large, in proportion to the size of the fruit, as that of the Guatemalan variety.

The pip of the West Indian fruit is more likely to 'take' and it usually grows more rapidly than the other.

Guatemalan

The skin of the Guatemalan fruit is green and smooth, although occasionally slightly pebbled. It weighs six to sixteen ounces. Its flesh is firmer and more watery than that of the West Indian fruit, and it has a lower oil content. The pip is large in proportion to the size of the fruit. Guatemalan avocados are much harder to start; they grow more slowly than the West Indian variety and they produce smaller – although charming – plants.

How to Judge Ripeness of Fruit

The easiest and fastest way to grow an avocado plant is to start with the seed from a ripe fruit. (It's better eating also.) There is one foolproof test for ripeness, even though it is less than popular with shopkeepers – the judicious application of a furtive thumb. Press at the stemmed end of the fruit. If it gives, even a bit, the fruit is fairly ripe. If it's soft, the fruit is fully ripe. But if it stonily resists all pressure (and is the only one to be had), take it and let it ripen at home. Don't keep it in the refrigerator. Put the fruit in a brown paper bag and set it aside. But keep an eye (and a thumb) on it: the fruit can ripen within a day or so.

If your shopkeeper tells you (as I've sometimes been told) that he prefers to carry the Guatemalan avocado because it doesn't turn black when it is opened, like the West Indian, I'd like to make a point: the flesh of *any* variety of avocado darkens when exposed to air for any length of time. But there is a remedy for this. If you plan to use the avocado in a salad very soon after opening it, a bit of lemon in the dressing will prevent the avocado from darkening and keep your salad green. If you are preparing a salad well before serving time, or are using a dressing without lemon, you can squeeze a little lemon juice on the avocado flesh to preserve its attractive colour. Store any cut-open avocado with a piece of grease-proof paper over the cut. If there are leftovers, try mashing a chunk of avocado into a hot, clear soup just before serving – it adds a delicious flavour.

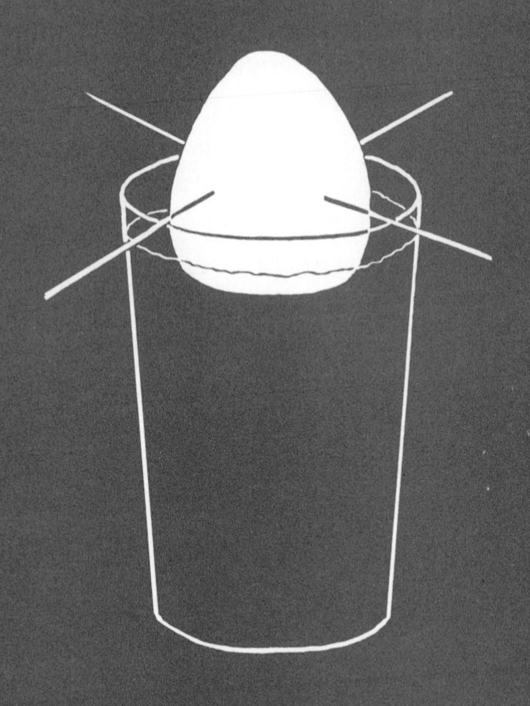

THE BASE OF THE PIP IN WATER

2

Starting the Seed

You'll need: A clean glass not less than five inches high and with a wide opening at the top, or a one-pound jam jar.

Wooden toothpicks. The rounded ones are stronger and easier to work with, especially if the seed is not very ripe. Metal hatpins are easy to use, too.

Identifying the Base of the Pip

The roots of the avocado plant grow out and down from the base of its big seed. (Once a plant has started to grow, that useless discard, the 'pip', can accurately be called a 'seed'.) But if you've never before looked at an avocado pip closely, you may be wondering which end is down. It is the place where a folded-in dimple can be seen at the bottom of the pip. Many pips are tapered up and away from a broad and clearly defined base. But some are quite round, or oval, with a flattened end. Look for the dimpled bottom if you're not quite sure which end of an avocado pip is the base.

Preparing the Seed

Once removed from the fruit, the pip may be a slippery object, wrinkled and pale, undivided and quite featureless, without much in the way of a coat. Or it may be roughly mottled, with a brown and distinctive papery coat. It may also be partially split, showing the first tendrils of growth clustered inside it, and the first roots putting out from the base. But no matter what its stage of development, it should be washed. However, *at no time should cold water be used on the avocado*. The plant is tropical in origin and will not take kindly to cold temperatures of any kind.

Washing the Pip

Rinse the pip in tepid water, removing as much skin as comes free easily. Don't dig into it. Be a little careful when handling a widely split-open pip. The two halves should not be entirely separated: if they are, a vital connective thread between them will be broken, and the seed's fertility may be impaired. But don't be too concerned about the way you handle the seed. By and large, it's tough and hardy, like the tree itself.

Preparing the Pip for Your Glass

Dry the pip and gently wipe it off, peeling away any stray shreds of skin that may still adhere to it. Put it aside. Fill your glass or jam jar with warmish water. Take up the pip again, and get a good grip on it. Then, about a third of the way up from the base, force half the length of four toothpicks into the pip. Place them at regular intervals, making a framework to support the pip across the top of the glass. (See page 16.) If the pip is very hard, or if you're having problems, use more toothpicks; they can't do any damage. Now place the toothpicked pip across the top of the glass, allowing about half an inch of water to cover the pip's base. Don't let the water come higher up or the pip may begin to mould.

Where to Keep Your Glass

The pip will now stay in water until it puts down roots. Place your glass in the warmest spot you can find, as you will need to maintain a temperature of 70°F. for the seed to germinate. Keep it out of strong light, natural or artificial. In dimness or shade the downward development of the roots will not be distracted by light from above. Your kitchen may be a very good place, warm and with water at hand, and perhaps a cupboard available to ensure good, dark shadows. A steamy warm bathroom is also a good place. Maintain a constant level of water in the glass at all times. The base of the pip should always be immersed. Make sure the pip is not exposed to town gas fumes, which are pure poison for most plants, including the avocado. However, natural gas is not poisonous and won't harm plants.

Watch for the first signs of activity to appear.

First Signs of Growth

It may only be a matter of a few days before the first roots appear, and when they do, your pip is now a full-fledged seed. But it may take longer to start rooting, so keep the seed where it is, and watch and wait. As long as the water in the glass stays clear, the seed is still sound. If you really get impatient, you can take the braced avocado seed from the water and look underneath it. You may see the first signs of life, a slender rootlet or two uncurling – or there may be no change in the seed's appearance at all. Be patient. Put it back in the water, back in the shade, and give it more time. I've had pips in water for as long as four weeks before they began to germinate. Remember that as long as the water remains clear your avocado seed is still healthy. Signs of decay will make themselves known soon enough. The water will begin to thicken and become cloudy, and a decaying pip has a rather unpleasant odour. Throw that pip away and start another. But if the pip was fertile to begin with, and if it is kept warm, wet and dimly lit, sooner or later it will certainly germinate.

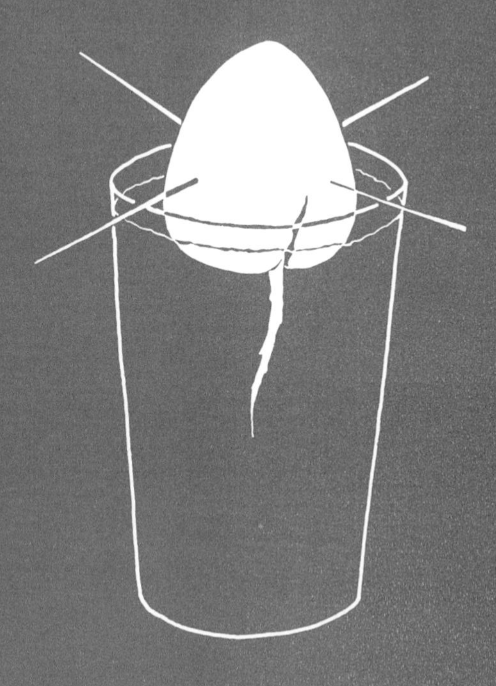

FIRST ROOTS

Second Signs of Growth

There are other signs of life to watch for, aside from the appearance of the first roots.

The seed will begin to split. If you started with a seed that had already clearly opened, you can be quite sure it's going to go on and become a plant. But if your seed is one of the tightly closed ones, its splitting is the next thing to notice. For me, this development is one of the most exciting parts of the whole process.

Splitting of the Seed

Bit by bit – and it can happen in a few hours or take as long as several weeks or more – the big seed eventually separates. There, lying in the centre of it, is the first fine, pale-green tendril, ready to shoot out and up into the air.

In some seeds there may be as many as three or four of these latent shoots in sight. They can start to grow rapidly, but one of them always takes the lead and will become the main stem or trunk of the plant.

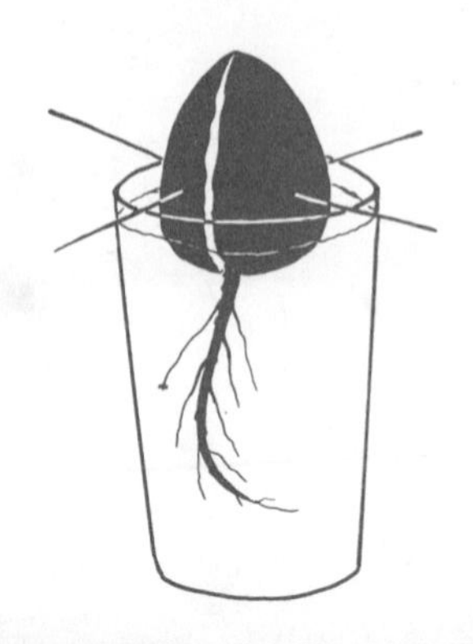

THE SEED BEGINNING TO SPLI

This main stem grows up and up, with the first tiny leaves held tight and closed at the tip of the green stalk. Or, in others, the reddish, freckled stalk may be only a few inches high when the first bronze-coloured, shiny leaves begin to uncurl. Obviously these are different varieties of the avocado. The concealed differences are only revealed when the seed begins to germinate and grow.

As soon as the first shoot emerges you should bring the seed into light so that it can begin to make its own food by photosynthesis.

Description of First Roots

A good growth of roots should develop with the new greenery. These roots are put out in several ways. They can be generous – falling thick and fast – and full, or they can be somewhat sparse – growing slowly – with a single root, thick, round and noduled. When both roots and stem develop at the same time, the seed has entered a state of good, active growth.

THE FIRST SHOOT

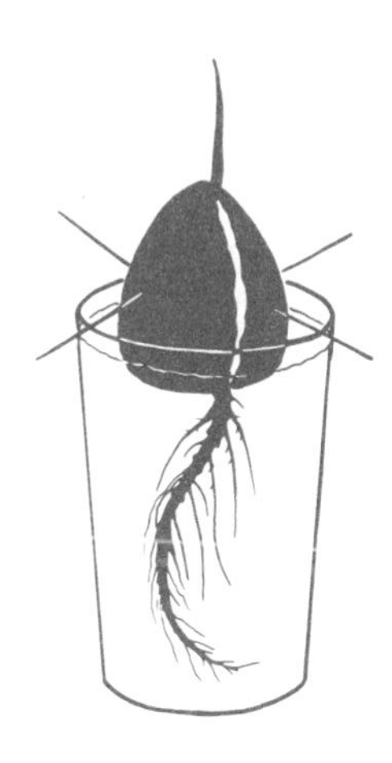

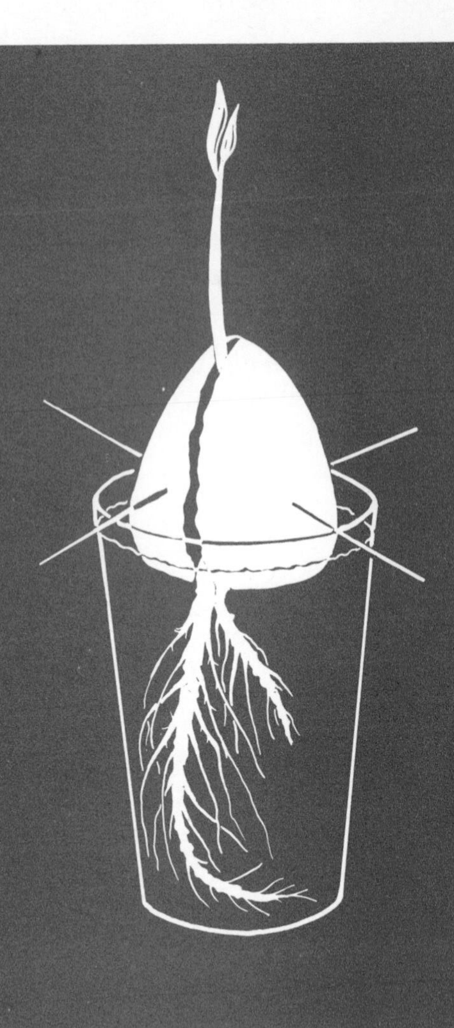

SHOOTS AND ROOTS, DEVELOPING TOGETHER

But an attempt must now be made to balance the development between the roots and the stem, and preference must be given to the roots. This is done by cutting back the stem, which serves to give the roots a better chance to put down while at the same time it checks the stem's too rapid growth.

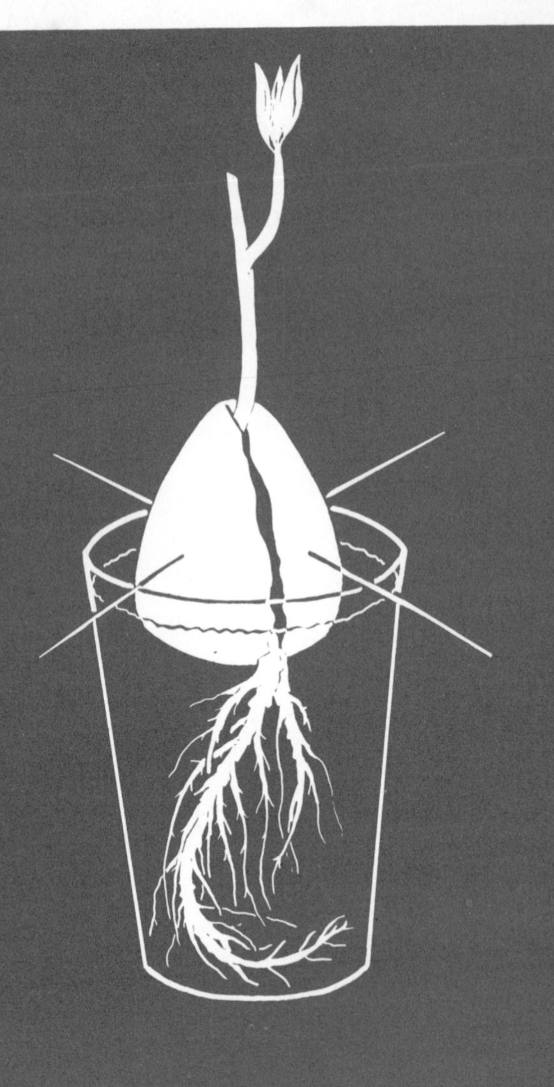

THE MAIN STEM, CUT BACK

3

Cutting Back While the Seed is in Water

You must try to be firm about cutting back an avocado's foliage. In their natural environment some trees grow straight up to forty feet. Others reach a more modest height of about twenty feet but are fuller and bushier. To achieve a many-branched and fully fruited tree, outdoor horticulturists control the tree's shape by cutting it back. The same rule applies to an indoor plant. If you want a lush, full, leafy but reasonably sized house plant, you must start cutting back at the very start. This cutting back is continued as the plant grows.

A seed may have several shoots growing up from the top. They look like smaller stems. Don't cut them back, just leave them alone. They will flourish, or not, when the seed is put into soil.

There is no mistaking the role of the main stem, however. It takes over at once, with great conviction. This slender column may already be tipped with leaves, opened or closed. Allow it to grow to a height of six or seven inches. Then, using a sharp knife, cut off the stem at a point midway between its top and bottom. Don't cut back *too* far. The remaining stem should be at least three inches high, but no shorter than that. By way of reassurance, let me explain that if you don't cut back the stem here and now, it will continue growing, on and on, producing leaves which will fall off, one by one, so that the plant, left to its own devices (in soil or water), will finally look like a telephone pole with only a few isolated leaves dangling from its tip.

The cut-off stem's development is now frozen. It may take another week or longer to start up again. Don't be impatient. The seed has to recover its resources, and this takes time. When the new shoot does appear it will grow out from somewhere along that three-inch stalk. Its growth will be slower, as will be the development of the leaves.

Time for the Growth of Roots and Stems

Roots. The first roots may appear in water in only a few days, or they may take four weeks or more to start growing.

Stem. The stem may appear in very little time, for in some very ripe seeds and in some varieties where the seed is already split open the stem can be seen before you lift the pip from the fruit. With other varieties, and depending on the ripeness of the fruit, it may be days or even several weeks before the stem emerges.

Even if you leave the seed in water for as long as six weeks you won't jeopardize the plant's future growth. Don't leave it too much longer, however. The seed can go bad if it's not put into soil within a reasonable time. The roots should be given space in soil before they look as if they might burst your glass.

When to Pot

You will be able to judge the time for potting in several ways.

1. A good, thick growth of roots is needed. They should be fairly dense and full, and long enough almost to reach the bottom of the glass. (But if the stem is growing extremely quickly, even after you've cut it back, and the roots remain only fairly thick, plant the seed immediately. It's a different variety of avocado from the kind that puts out the slender but quicker-growing roots.)

2. You should plan to plant the seed about two weeks after you cut back the stem. It can be left longer, but don't postpone the transfer for more than three weeks.

It is desirable to have a good root system well established before you plant. What is needed is to get the plant into soil fairly soon so that it can start producing its own nourishment from the light and soil, rather than from its seed.

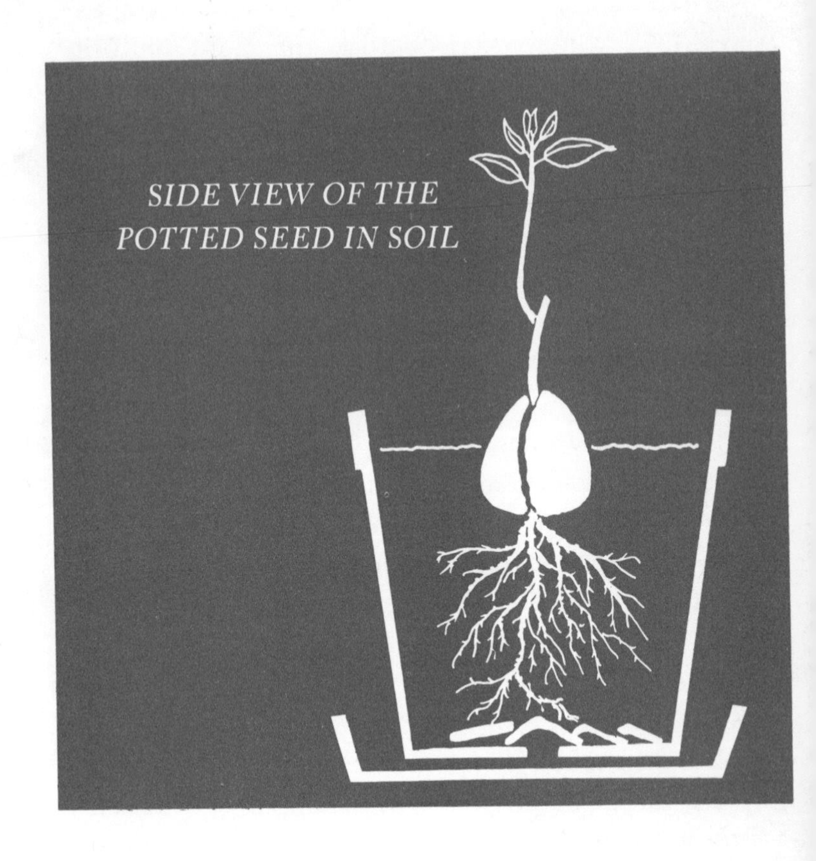

4

Planting the Seed in Soil

You'll need: 1. Flower pot. 2. Dish to go under pot. 3. Broken crockery, or an expendable clay flower pot, broken. 4. Soil. 5. Bamboo cane or stick.

Notes:

1. *Flower pot.* The inside diameter of the top of the pot should be five inches.

The flower pot you use is of prime importance to your avocado's welfare. The avocado's roots must have free soil drainage at all times. Avoid so-called 'self-watering' pots: they do not give a sufficient flow of water. Get the familiar kind of inexpensive clay pot that sweats. Avoid painted pots: they tend to hold moisture.

2. *Dish to go under pot.* Don't use a clay dish. It oozes and will water-stain the surface of whatever it is placed on. Avoid metal pans. Unless they're of

high-grade stainless steel or aluminium, they rust, look messy and finally fall apart. A pyrex glass dish is a good vessel for holding water and can be cleaned easily.

3. *Broken crockery or an expendable or broken clay flower pot*. These are to be used, broken-up, placed around and over the drainage hole in the bottom of the pot before you add the soil. This prevents too dense packing of the root ball. Fold a sheet of newspaper over the crockery, and using a hammer, break it into chunks. Don't crush or break into too small pieces.

4. *Soil*. The avocado will grow best in a rich loamy soil that is well drained. A good compost to use is a John Innes mixture, No. 1, for the first potting as this contains sufficient nutrients for the young plant in its early stages of growth. Bags of John Innes compost are available from shops selling gardening supplies.

If you choose to use a loamless compost, or one that does not contain nutrients, then you must be more careful about feeding the plant adequately and you will need to begin to feed it earlier than you would with a John Innes compost.

Never use any soil dug up from your garden: it may not contain sufficient nutrients and it will probably be carrying pests and diseases that will harm your avocado.

5. *Bamboo cane or stick*. One at least three feet tall.

Another up to eight feet, if you're planning to grow a tree.

The plant's stem should have the support of the cane as soon as it reaches a height of sixteen or seventeen inches. If your plant turns out to be tall and thin-stalked, the cane will have to be put alongside the stem four weeks after planting the seed.

You can buy sticks at most supply stores, painted and fairly expensive. Or you can buy a ten- or twelve-foot length of cane at any gardening shop for much less and cut it into the lengths you need. You can colour it yourself so that you will have a stick with a matt finish which you can cut into a three- or four-foot length for the first planting and have a piece left for later needs. Don't use oil paints or chemical stains.

Potting

If you are using a clay pot soak it in water for a couple of hours before potting – otherwise it will absorb moisture from the plant's compost.

Take the seed out of the glass or jam jar and remove the toothpicks. If they can't be pulled free then gently break them off close to the seed. Put the seed to one side, taking care not to damage the young roots and shoots. Arrange the broken crockery at the bottom of the pot around the drainage hole. Cover the crockery with soil, but don't firm it down.

The avocado seed should not be fully covered with soil when planted. Hold the seed with one hand and lower it into the pot so that the top of the seed where it is split is level with the top rim of the pot. Sprinkle soil around the roots with your free hand, gently moving the seed about as you do this to ensure that soil gets between all the roots. Carry on filling the pot until the soil is halfway up the seed – about half an inch below the top rim of the pot. Firm the soil so that the plant is held securely – but don't compact it – and then tap the pot firmly on the table to settle in the plant and soil.

Soak the soil with tepid or warm water from above, then stand the pot somewhere so that the excess water can drain away freely.

Support the plant by forcing a cane six to twelve inches taller than the plant into the soil close to it. Tie in the plant, firmly but not too tightly, with green gardening twine.

Where and How to Keep the Plant

Now that the young main stem has been cut back (see page 27), the plant's aerial growth is at a temporary stand-still. This can be an uneasy period, while you wait and watch; it may be weeks before the stem starts up again. However, a new but shorter

THE PLANT GROWING IN SOIL

stalk will finally put out from the truncated stem. The new stalk will appear sooner if the plant is placed in strong light.

From now on, your plant should have as much light as you can give it.

The most convenient source is sunlight. If there's a spot in your house that enjoys several hours of sunlight a day, this is the best place for your plant. But the plant will respond just as effectively to artificial light, provided it's strong enough and is supplied often enough.

The direct light from two 100-watt bulbs for a few hours a day or more is adequate. Don't burn the leaves by letting them touch the light bulbs.

To ensure an even, constant distribution of light, change the plant's position from time to time, turning it in the light.

Several hours of light from a frosted white fluorescent tube is excellent, second only to sunlight.

Temperatures

Cold. Anything under 36 degrees Fahrenheit is risky for an avocado. The Guatemalan (California) variety may be able to withstand a little frost, but try to avoid extreme chilling. Guard your avocado against wintry blasts from open windows.

Heat. The plant will thrive in any warm temperature, providing the air is humid. During the summer months the plant can be put outdoors. If you have a balcony or patio, the plant can be kept there all summer.

Your plant can be kept on top of a radiator. But be very certain that there is a heavy or asbestos cover over the hot pipes. If heat becomes too extreme or too direct from below, the plant can be cooked, which is a bad thing. I keep a number of plants above a radiator that has an asbestos protective layer over its surface. Draughts of hot air stir the plant's leaves, and they flourish and grow, perhaps because there is such a tropical atmosphere above, as well as blowing up from below.

Watering

For your plant to thrive it is important that, as far as possible, you try to re-create its natural habitat. The avocado is a tropical plant, and, as well as being warm, the tropics are very humid, so you must keep your plant moist.

Plants lose moisture to the air by a process called transpiration, which is something a lot more complicated than, but similar to, evaporation. They will transpire more when the air is warm and dry and there is a breeze – ideal conditions for drying out washing! The amount of water that you give to your plant via its roots will depend on how large the plant is, whether it is dormant or growing vigorously, how warm the temperature is and how humid the atmosphere. Be very careful not to over-water your plant via its root system. If the soil becomes water-logged there will be no space left in the soil for air, and if the roots are unable to breathe they will die – as will your avocado plant. Plants are being over-watered if they droop and flag while the compost is still wet, and if leaves yellow and drop.

Water your avocado *little* and *often*, daily if necessary. Water it from above, and then if a lot of excess water seeps through to the saucer below throw it away – don't allow the plant to stand in this water.

In watering your plant through the roots you are replacing the water that it is losing by transpiration. It is best to try to create an environment which will prevent the plant from transpiring excessively. As your avocado needs warmth you can't lower the temperature, and so you must try to keep the atmosphere around your plant as humid as possible. Do this by thoroughly spraying your plant every day – more frequently if the weather is warm. Washing the leaves with a sponge will help and it will also remove dirt and grime, but never use soap!

Always remember to use *warm* water!

A gardening note. After a while the surface of the soil becomes packed rather solidly in the pot. When this happens take a fork and gently loosen the soil. Turn it over, digging not more than two or three inches below the surface, being careful not to damage the roots. This serves to aerate the soil, and is beneficial for the plant.

5

Cutting Back the Young Plant

The Young Plant

You have already cut back your plant. The main stem was shortened while the seed was still in water. No more cutting back is required now.

About the time the main stem puts out its first shoots of leaves, a secondary stem may have begun to emerge from the base of the seed (or it may have been visible from the very beginning). Don't cut back that secondary stem. It can grow up and become an adjunct to the main stem, putting out its own leaves, and making a double-trunked plant. A seed may show more than one of these tiny, tentative stems, and at times even a third or fourth may be seen. Leave them alone. The second stem will grow or it won't, while the remaining stalks often don't come to much. Wait and watch.

The Mature Plant

The avocado is a tree and the main stem will become the trunk of that tree. From that parent trunk stems will grow, and will in their turn become branches. And from those branches still other stems will grow. And all of these stems and branches will be leafed.

Pruning keeps a plant's shape within attractive and practical bounds. But pruning also serves to stimulate the growth of dormant stems and branches that are prevented from developing because of too vigorous and rapid *outward* and *upward* growth.

The young branches growing from the *bottom* of the parent trunk should be encouraged. To do this, prune any stems that are growing on the upper branches. Go on in this fashion, always working down from the top of the plant. Prune all younger stems and shoots of leaves that give a lopsided look to the plant as a whole. Some of these stragglers are temptingly green. But they have to go if you want your plant to be fully stemmed and leafed, with a symmetrical, *contained* shape.

Don't be afraid of pruning. You can't do the plant any serious damage by removing a crown of leaves or even whole and fully leafed stems. The avocado is tough and grows rapidly. Be bold; experiment with it. And watch. Successful growers all started by being good watchers.

Use secateurs, or a sharp gardening knife. All

WHERE TO CUT BACK THE MATURING PLANT

branch cuts should be clean ones. Cut close to the trunk and branch so no unsightly stubs are left. You can pinch out single leaves and crowns with your fingers. Pinching out is merely another form of pruning.

THE HEALTHY PLANT, CUT BACK FOR SYMMETRY

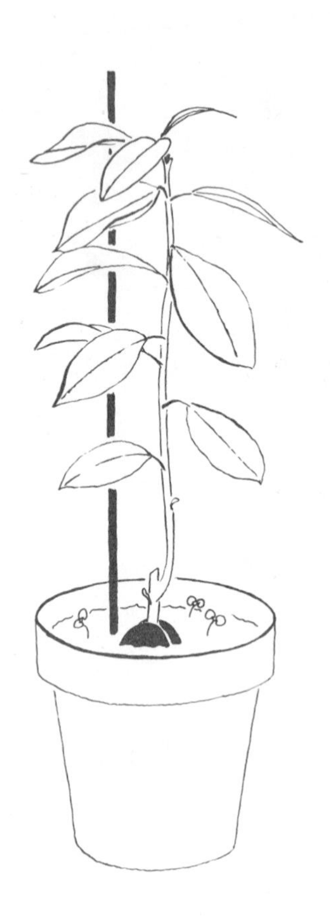

Remember to prune from the top and the outside. That way you will force the plant's growth down, into, and towards the already existing greenery.

The Tree

Continue to apply the above.

The avocado has a resilient branch that dips and curves down as it grows out and away from the trunk. Sometimes one of these branches will reach out until it is nothing more than a bare stretch of leaf-tipped stem. If you cut off the whole thing you will soon see how the other branches benefit from its removal. New stems with budded leaves appear in unexpected places, on lower, older branches as well as on other stems and branches.

THE TREE TRANSPLANTED TO A WOODEN TUB

6

Transplanting into a Larger Pot or Tub

Your avocado plant will probably need to be re-potted from its first five-inch pot after two months. You should move it to a ten-inch pot, or one even larger, as this will be its permanent home.

Signs that the plant needs re-potting are roots appearing through the drainage hole or on the soil surface, leaves yellowing and dropping even if you've just watered it, or the plant just not developing any new growth although it should be doing so.

You'll need: 1. Large pot or tub – make sure that it has a drainage hole. 2. Dish to go under the pot. 3. Soil. 4. Broken crockery. 5. Cane to support the plant.

1. *Large pot or tub.* The pot should be at least ten inches in diameter at its top rim, as this will be the plant's permanent home. Choose one that is attractive, but if you decide to use a large wooden tub

instead of a clay pot make sure that it has a drainage hole.

2. *Dish.* You can usually buy clay dishes to go with clay pots, or use an unwanted pie dish, or something similar.

3. *Soil.* The soil is very important as your plant will be growing in it for some time. John Innes No. 3 compost is ideal as it will supply the nutrients that your plant requires for up to a year, but you may use a loamless compost if you wish. If the compost that you choose contains plant food then begin to feed your avocado after three to six months; if the compost is without nutrients feed it immediately. All plant composts should state quite clearly on the bag what they contain – if they don't then you should not buy them.

4. *Pieces of broken crockery.* Use for drainage.

5. *Cane.* This should be one to two feet taller than the height of the plant. A strong bamboo cane available from any shop selling gardening supplies will be fine.

Re-potting

If the pot you are using is made of clay soak it thoroughly for a few hours to prevent it taking moisture from the plant's compost.

To remove the plant from its old pot hold it upside-down, your palm covering the soil with the plant protruding downwards between two fingers. Now firmly knock the pot rim against a table edge to loosen it. Gently pull the pot away, taking care not to damage the roots.

Fill the bottom of the new pot with a few pieces of broken crockery and cover these with compost. Hold the plant in position in the pot – the top of the root ball should be half an inch below the top rim. Fill the pot with compost, firming it with your fingers as you go, but don't compact it. Cover all the roots, bringing the compost up to half an inch below the pot rim. Tap the pot firmly on the table to settle the soil. Water thoroughly and leave somewhere for the excess water to drain away.

Re-potting is best carried out in the spring or early summer as the daylight and weather conditions will encourage the plant to grow quickly and fill out its new pot. However, if your plant needs to be re-potted and it's the middle of winter then you should go ahead and pot it up as it will harm the plant to be kept in cramped conditions.

Your avocado will need to be supported. Push the cane into the pot near to the stem and, using green gardening twine, tie in the plant firmly, but not so the string will cut into the plant.

Your plant can now be moved to its permanent home.

Light for the Tree

The tree should have lots of light. If you don't have sunlight, try to arrange for some sort of artificial light that will fall on the leaves from top to bottom for as many hours as can be managed. The more light you give the tree, the better it will grow. A floodlight would be ideal.

Turn the tree around from time to time. The shape of your tree's growth and development is affected by the directions from which light reaches it.

The Best Time to Grow or Start Plants

I used to think that avocados grew better if started in the spring of the year. I've long since given up the idea. The plant can be started, and will grow well, at any time of the year. It only needs to be kept warm and moist.

7

General Care of Your Plants

Avocados are strong and healthy plants. Once established your avocado should thrive without setback. The great majority of disorders that your avocado is likely to encounter are caused by incorrect care and attention. Be observant. Watch for dead, dying and mottled leaves and respond quickly. Remember that prevention is better than cure, and in many cases it is only when plants are neglected that they become prone to infection. So let us start with general care.

Watering

Probably more house plants die from incorrect watering than from any other single cause. The symptoms of over-watering and under-watering are the same because both produce the same result: dehydration of the leaves.

In the case of under-watering dehydration is a

fairly obvious result. In the case of over-watering it needs some explanation. To grow and function effectively roots must have water and air. Over-watering, however, causes all the air to be driven out of the plant's soil, depriving the roots of oxygen, which they need to breathe. Even without oxygen, the roots will continue to respire and in so doing they produce alcohol as a by-product. Alcohol is highly toxic to plants and quickly kills the roots. Dead roots are naturally unable to perform their function of supplying the leaves with water, and so the leaves died from dehydration.

The symptoms of both over- and under-watering are, therefore, the same. In both cases leaves wilt and eventually drop, and the leaf edges may turn brown. Growth will be poor and stunted.

If your avocado is badly under-watered you may be able to revive it by a thorough soaking. Place the plant in a deep container filled with tepid water so that the water covers the top of the soil. Air bubbles will emerge from the soil. The plant is soaked when no more bubbles emerge and you can then remove it and leave it to drain and recover.

A severe case of over-watering may be helped by removing the plant from its pot and leaving it to dry out for 24 to 48 hours. After returning the plant to its pot it's a good idea to water it with a fungicide solution such as Benomyl as the decayed and rotted roots will be a wonderful breeding ground for fungal infections. Water your avocado more judiciously in future and remember to keep the air humid.

If you have to leave your plant unattended for some time plunge the pot in a large container filled with moist peat and remove the plant to a slightly cooler and darker place; both these procedures will help to lower the plant's loss of water by transpiration.

Temperature

You may find that your avocado suddenly drops a great number of leaves; it can, in fact, lose up to a third of its leaves overnight. This is caused by a severe shock to your plant's system, usually as a result of a radical change in temperature over a short period. This is obviously more of a problem in winter when the temperature in your home will be higher in the day and may drop quite severely at night. An avocado objects more to a change in temperature than to a steadily maintained cooler environment. It is far better to aim for a stable temperature of about 60°F. than one of 75°F. that will fall when the heating is lowered. Be careful to keep your avocado out of the way of any draughts as they can cause wide and sudden fluctuation of temperature.

Light

Small pale leaves with weak, elongated stems are a sure sign that your avocado is not getting sufficient light. Increase the availability of light, if necessary by using daylight (not white) fluorescent tubes.

Lop-sided growth is an indication that the light is unevenly distributed. To rectify this, give the plant a quarter of a turn every day.

Under-potting

If your avocado is growing slowly, even when fed, and its compost dries out very quickly, look for the appearance of roots through the drainage hole and on the surface of the soil. Your avocado is probably pot-bound and should be re-potted into a larger container immediately.

Feeding

Plants manufacture food by a process called photosynthesis. In photosynthesis they make carbohydrates from carbon dioxide (absorbed from air) and water (absorbed through their roots) with the energy they receive from light, which is why light is so vital to all green plants. For photosynthesis to take place, as well as the other numerous processes by which plants grow and reproduce, they must have a wide variety of minerals and trace elements. These nutrients are usually available in the soil in which the plant is growing. However, the amount of soil in a pot is obviously limited, so that sooner or later the foods will be used up, or washed out of the soil by constant watering. You must, therefore, feed your plant.

The amount of food a plant needs naturally depends on its size and on how rapidly it is growing. Avocados, like other plants, have resting periods, which, to you, may seem quite arbitrary. These dormant periods depend on the variety of the plant and where it originally came from. Respect your avocado's resting period: do not feed it if it is obviously not producing new growth, but watch for signs of growth and begin to feed it immediately.

Use a liquid food. These are available from all shops selling gardening supplies. Make sure that it is a balanced food containing a variety of essential nutrients – not a food to remedy specific mineral deficiency. Liquid plant food is highly concentrated and should never be used undiluted. Nor should you use too much; it will not help your plant to grow any faster and could seriously damage it.

So, watch your plant for signs that it needs food – new growth, for example, or yellowing, mottled leaves.

If you planted your avocado in a loam-based compost containing plant nutrients, such as the John

Innes No. 1 used for its first potting, you won't need to feed it for four to six months – by which time you should have re-potted it anyway. John Innes No. 2 will feed your plant for six to nine months, and John Innes No. 3 for up to a year. After these periods, or if the compost you've used is loamless and doesn't contain nutrients, feed the plant every two weeks while it is growing.

Fungal Diseases and Hygiene

Many fungal diseases of plants occur as a result of poor hygiene. To reduce the risk of infection from these diseases it is important that you keep the soil, the pot and the plant itself clean.

Always use fresh, sterile soil. This is available in plastic bags from a gardening supplies shop. Never use soil from your garden, or old soil from another plant: it will be lacking in minerals and may act as host to a variety of fungal infections and pests.

Wash the pot that you are using very thoroughly. Plastic pots are easier to clean than clay ones, but a scrub with plenty of hot water should suffice.

Remove and throw away any dead and dying leaves. If you allow them to remain on the plant or leave them to rot on the soil they may harbour fungal diseases. Leaves that have become mottled or damaged will not recover, and your avocado is better off without them. Take out any weeds in the pot; they frequently act as hosts and breeding grounds for pests.

Pests

There are a small number of pests that could infest your avocado. They all can be dealt with easily and effectively by using suitable chemicals in the early stages of attack.

Red Spider Mite and thrips will produce pale or brownish mottling, particularly on young leaves. With a severe infestation of Red Spider Mite a mass of fine webs will appear on the underside of affected leaves. New growth will be damaged and deformed.

To clear the pest spray with Liquid Derris or Malathion as directed by the manufacturer. Isolate the infected plant from others as the pests can spread rapidly from plant to plant.

Aphids are tiny green insects which attack the young growth of plants by stabbing the tissue with their sharp mouth-pieces. This damages and deforms new growth giving it an unsightly appearance. Aphids excrete honeydew, an unpleasant sticky substance that encourages the development of Sooty Mould, a black-coloured fungal disease.

Aphids can be killed with Pyrethrum or Pirimicarb. If Sooty Mould has developed wipe the affected leaves with a mild detergent solution and then again with clean water.

A bad attack of *White Fly* is hard to miss as clouds of small white insects fly up into the air whenever the leaves are disturbed. It is the White Fly's larvae and not the more visible adult fly that harm the plant. They stick firmly to the underside of the leaves where they proceed to suck the plant's sap. White Fly can be difficult to control because of their astonishingly rapid rate of breeding.

To eradicate White Fly spray with Pyrethrum, Resmethrin or Malathion for four weeks.

All the above-mentioned chemicals are toxic and unpleasant but they are not at all dangerous if used sensibly, in accordance with the manufacturers' recommendations.

A FRUIT-BEARING, FULL-GROWN TREE, OUTDOORS

8

The Botany of the Avocado

The botanical name of the avocado is *Persea grat-issima*, although sometimes it's known as *Persea americana*. It is a member of the plant family Lauraceae.

Family

The family Lauraceae is made up of evergreen trees and shrubs found in tropical and sub-tropical regions, many of which are used for timber production. Other members of the Lauraceae family are the cinnamon tree (*Cinnamomium zelancium*) from Sri Lanka, the camphor plant (*Cinnamomium camphora*) found in China and Japan, the sassafras plant (*Sassafras officinale*), a native of North America; and the bay tree (*Laurus nobilis*) which is commonly grown as an ornamental tree or shrub.

Genus

The genus *Persea* contains 150 species, all of which are trees found in the tropics.

The red or sweet bay (*P. borbonia*) yields fine wood used for making furniture. The bark of two other species (*P. lingue* and *P. meyeniana*) is used for tanning.

There are two other species of *Persea* which produce edible fruits similar to the avocado pear which we eat. These are the coyo avocado (*P. schiedeana*), which has small fruits that are eaten locally, in Central America, where the tree grows, and the Mexican avocado (*P. drymnifolia*). The latter is frequently, although incorrectly, thought to be a variety of the avocado pear, but its fruits are smaller with thinner skins and its leaves are highly aromatic. The Mexican avocado is often used as a root-stock for the Guatemalan race of avocado pear when it is grown commercially.

Species

The avocado pear species, *Persea gratissima*, contains two races or varieties: the Guatemalan and the West Indian.

The Guatemalan variety is native to the highlands of Central America. Its fruit is variable in size, from ten to thirty-five ounces, and can be pear-shaped to flattened-oval. The fruit takes nine to twelve months to ripen on the tree.

The West Indian variety is less hardy (less frost-resistant) than the Guatemalan. The fruit is large and can weigh up to three pounds. Its skin is thick, leathery, pliable and coloured dark purple. The fruit takes six to nine months to ripen on the tree.

The leaves of the avocado pear tree are evergreen, elliptic or ovate and from fourteen to sixteen inches long. The wood is very light and brittle.

Pollination

The avocado pear's flowers are borne in 'racemes' – similar to lilac blossom. They measure about half an inch across and are coloured a pale yellowish-green. The plant is bisexual: the flowers contain both the male and female parts within them.

For the avocado to reproduce and therefore to bear fruit, the male gamete, which is the pollen, has to fertilize the female gamete, the ovum. The stigma – the female part of the plant which receives the pollen – usually matures before the pollen is ready to be produced within the same individual flower. For fertilization to occur the pollen must reach the stigma of another flower that is able to accept it and which will therefore be at an earlier stage of maturity. Pollen is usually carried to the stigma by insects such

as bees, although this task can be carried out manually. It is very rare for avocados grown indoors to produce flowers, but if your avocado does flower then to get it to bear fruit, try pollinating manually – you never know!